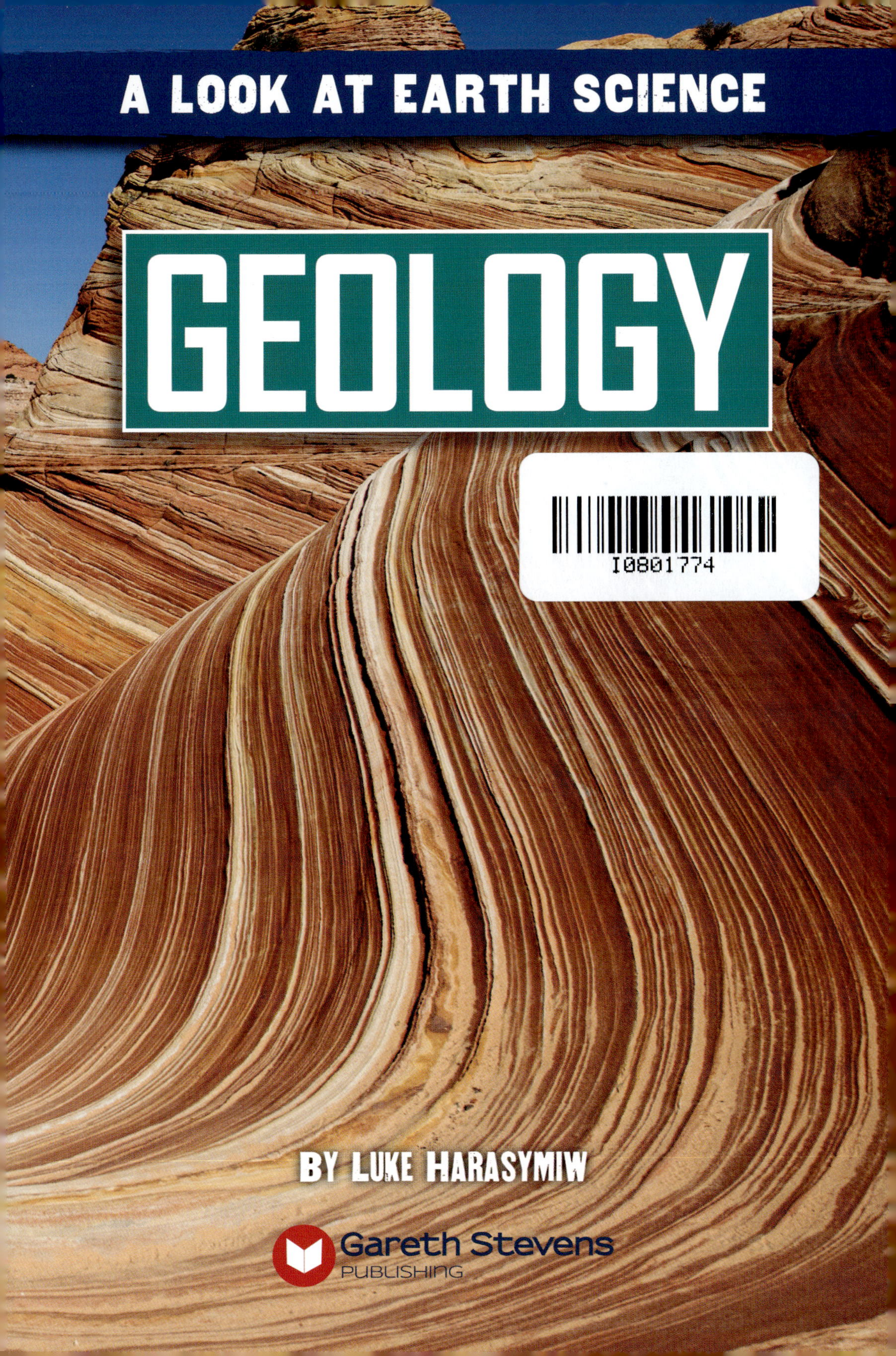

A LOOK AT EARTH SCIENCE

GEOLOGY

BY LUKE HARASYMIW

Gareth Stevens PUBLISHING

Please visit our website, www.garethstevens.com. For a free color catalog of all our high-quality books, call toll free 1-800-542-2595 or fax 1-877-542-2596.

Cataloging-in-Publication Data

Names: Harasymiw, Luke.
Title: Geology / Luke Harasymiw.
Description: Buffalo, NY : Gareth Stevens Publishing, 2025. | Series: A look at Earth science | Includes glossary and index.
Identifiers: ISBN 9781482467208 (pbk.) | ISBN 9781482467215 (library bound) | ISBN 9781482467222 (ebook)
Subjects: LCSH: Geology–Juvenile literature. | Petrology–Juvenile literature.
Classification: LCC QE29.H375 2025 | DDC 550–dc23

First Edition

Published in 2025 by
Gareth Stevens Publishing
2544 Clinton Street
Buffalo, NY 14224

Designer: Jennifer Schoembs
Editor: Therese Shea

Photo credits: Cover, p. 1 defmode11/Shutterstock.com; series art (backgrounds) KADIVAR07/Shutterstock.com; p. 5 beboy/Shutterstock.com; p. 7 Dima Zel/Shutterstock.com; p. 9 thoron/Shutterstock.com; p. 11 (top) Mike Beauregard/commons.wikimedia.org; p. 11 (bottom) Gerald Corsi/iStock.com; p. 13 Stephencdickson/commons.wikimedia.org; p. 15 Ali DM/Shutterstock.com; p. 17 Havardtl/commons.wikimedia.org; p. 19 Kevmin/commons.wikimedia.org; p. 21 Peter Hermes Furian/Shutterstock.com; p. 21 (background) Semnic /Shutterstock.com; p. 23 (bottom) Rainer Lesniewski/Shutterstock.com; p. 23 (top) Wend Images/Shutterstock.com; p. 25 Adwo/Shutterstock.com; p. 27 IrinaK/Shutterstock.com; p. 29 Pornpimon Ainkaew/Shutterstock.com; p. 30 (icons) Icongeek26/Shutterstock.com.

Printed in the United States of America

CPSIA compliance information: Batch #CS25GS: For further information contact Gareth Stevens at 1-800-542-2595.

CONTENTS

Words in the glossary appear in **bold** type the first time they are used in the text.

STUDY OF OUR PLANET

Geology is the study of Earth. That includes what Earth is made of, how it formed, and how it has changed over time. It also includes the **processes** and forces still acting on Earth today. Earth is always changing, sometimes slowly and sometimes quickly.

MAKE THE GRADE

A volcanic eruption is the bursting forth of hot, liquid rock from within Earth. It can change Earth's surface, or top **layer**, very quickly.

EARTH DETECTIVES

Scientists who study geology are called geologists. They're like detectives looking for clues about Earth's history. Much of that history is hidden deep within Earth. Earth's different processes can uncover clues, however. Rocks contain some of the best clues about Earth's past.

MAKE THE GRADE

Earth is about 4.5 billion years old!

RECORDED BY ROCKS

Earth's rocks are made up of **minerals**. Geologists study these minerals closely. The rocks can reveal, or tell, if water once covered a place. They can reveal facts about **climates** of long ago. They can tell what gases were in the air back then as well.

MAKE THE GRADE

Rock may form in layers called strata. Some geologists study strata, a field called stratigraphy.

Geologists can use knowledge of **elements** in rocks to find out how old they are. For example, certain bits of matter break down over a known period of time. Figuring out how much of this matter has broken down in all can reveal a rock's age.

MAKE THE GRADE

Canada is home to the oldest rocks discovered so far. The rocks shown here, found in the Acasta River area in the Northwest Territories, are about 4 billion years old.

THE ROCK CYCLE

Some geological changes on Earth are slow. We can't see them happening. Others are quick and easy to spot, such as **earthquakes.** The rock **cycle** describes the slow and quick processes that cause Earth's three main kinds of rock to form and change.

MAKE THE GRADE

James Hutton (1726–1797) is called the founder of modern geology. The rock cycle was one of his ideas. He believed rocks were always changing form.

These are the main parts of the rock cycle: Rock is worn down through **weathering** and **erosion**. Bits called sediment settle and pack together to form sedimentary rock. Deeply buried rock melts into magma, or liquid rock. When magma cools, it forms igneous rock. Under great heat and **pressure**, metamorphic rock forms.

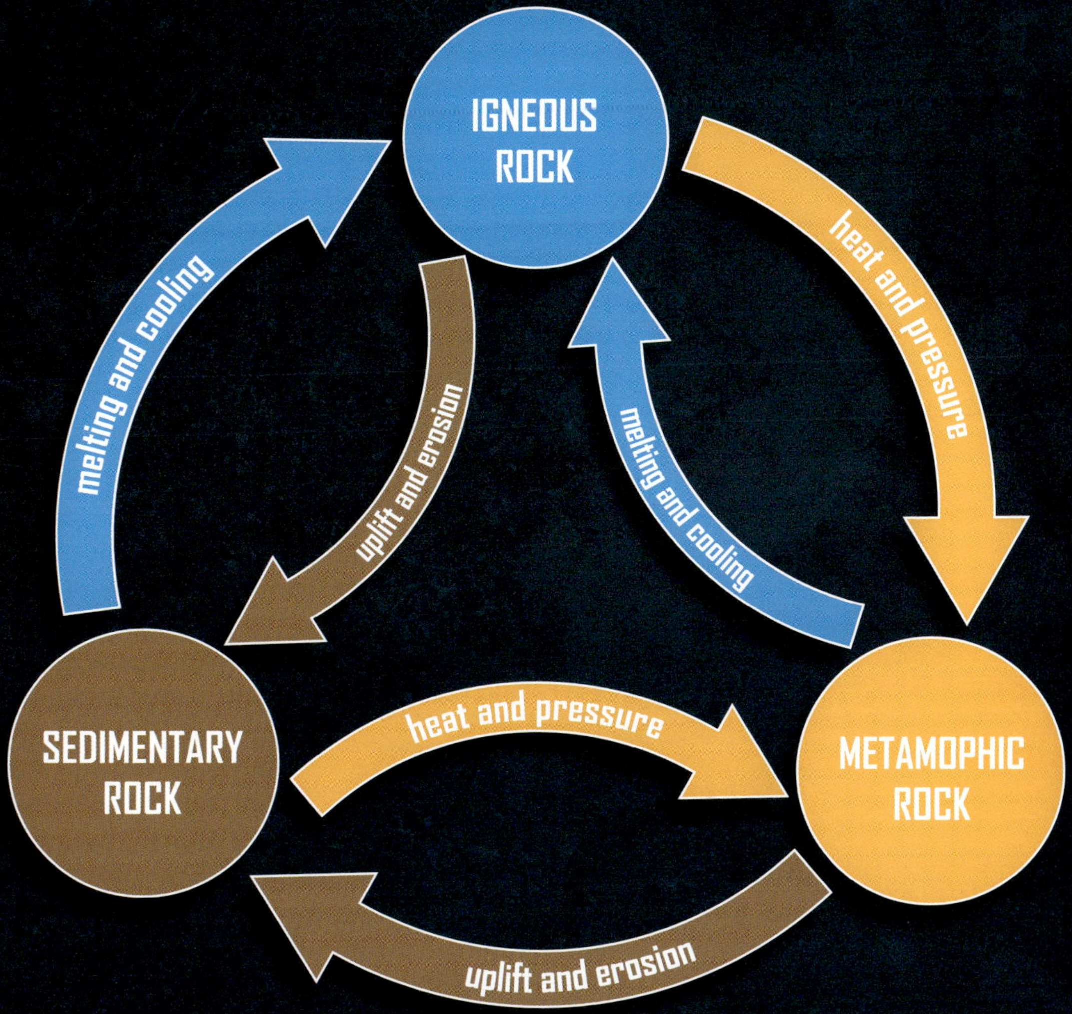

MAKE THE GRADE

The rock cycle doesn't follow any certain order. The different types of rock can take hundreds, thousands, or millions of years to form.

FOSSILS

Fossils are marks or remains of plants and animals from thousands or millions of years ago. They're often found in sedimentary rock. Bits of sediment form layers over the plant or animal remains. This is how some fossils keep their form until uncovered.

dinosaur footprints, Bolivia

MAKE THE GRADE

Trace fossils are evidence, or clues, left by living things that show how they lived, such as footprints and plant roots. Body fossils are evidence of what living things' body parts looked like.

only during certain time periods. If their fossils are found in a rock, that rock is from that time period.

MAKE THE GRADE

Fossils provide evidence that **continents** that are far apart now were once connected! For example, fossils of the **reptile** *Mesosaurus* have been found in South America and Africa.

Mesosaurus, *Brazil*

PLATE TECTONICS

Plate tectonics is the idea that Earth's outer layer is made up of large, slowly moving pieces called plates. Most geologists accept plate tectonics as fact. Plate tectonics helps explain certain geological events. For example, when tectonic plates come together or move apart earthquakes and volcanoes may happen.

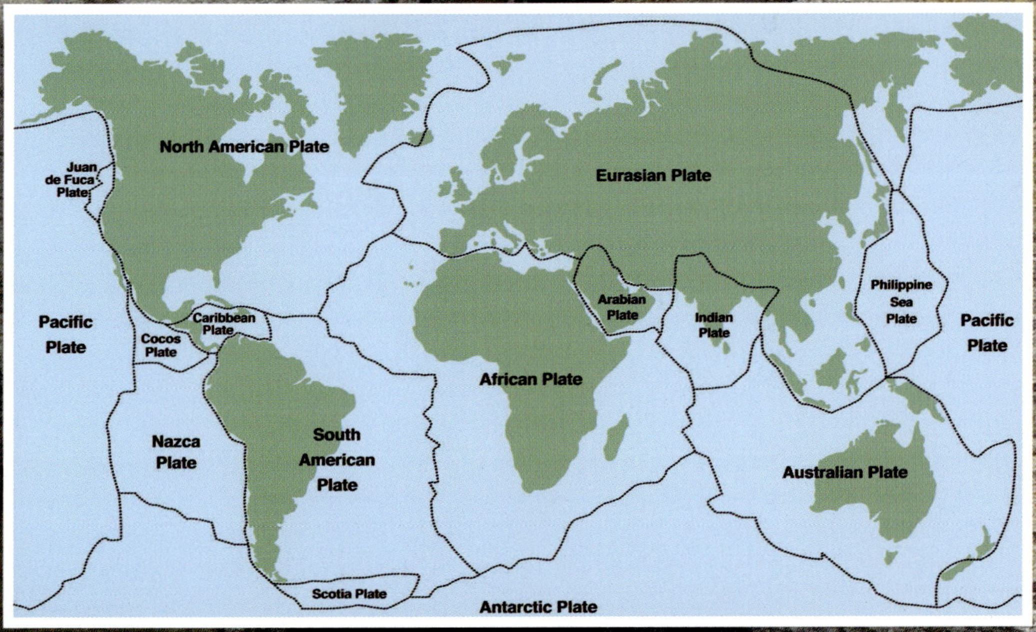

MAKE THE GRADE

Two tectonic plates pushing together may build mountains. Mount Everest, the tallest mountain on Earth, is getting taller each year as the Indian Plate is slipping under the Eurasian Plate.

Many volcanoes and earthquakes happen around the edges of the Pacific Plate. This area, shaped like a horseshoe, is called the Ring of Fire. In the United States, the cities of Seattle, Washington; Portland, Oregon; and San Francisco, California, are in the Ring of Fire.

MAKE THE GRADE

Geologists who study earthquakes are called seismologists. Those who study volcanoes are called volcanologists. Famous volcanoes in the Ring of Fire are shown in this map.

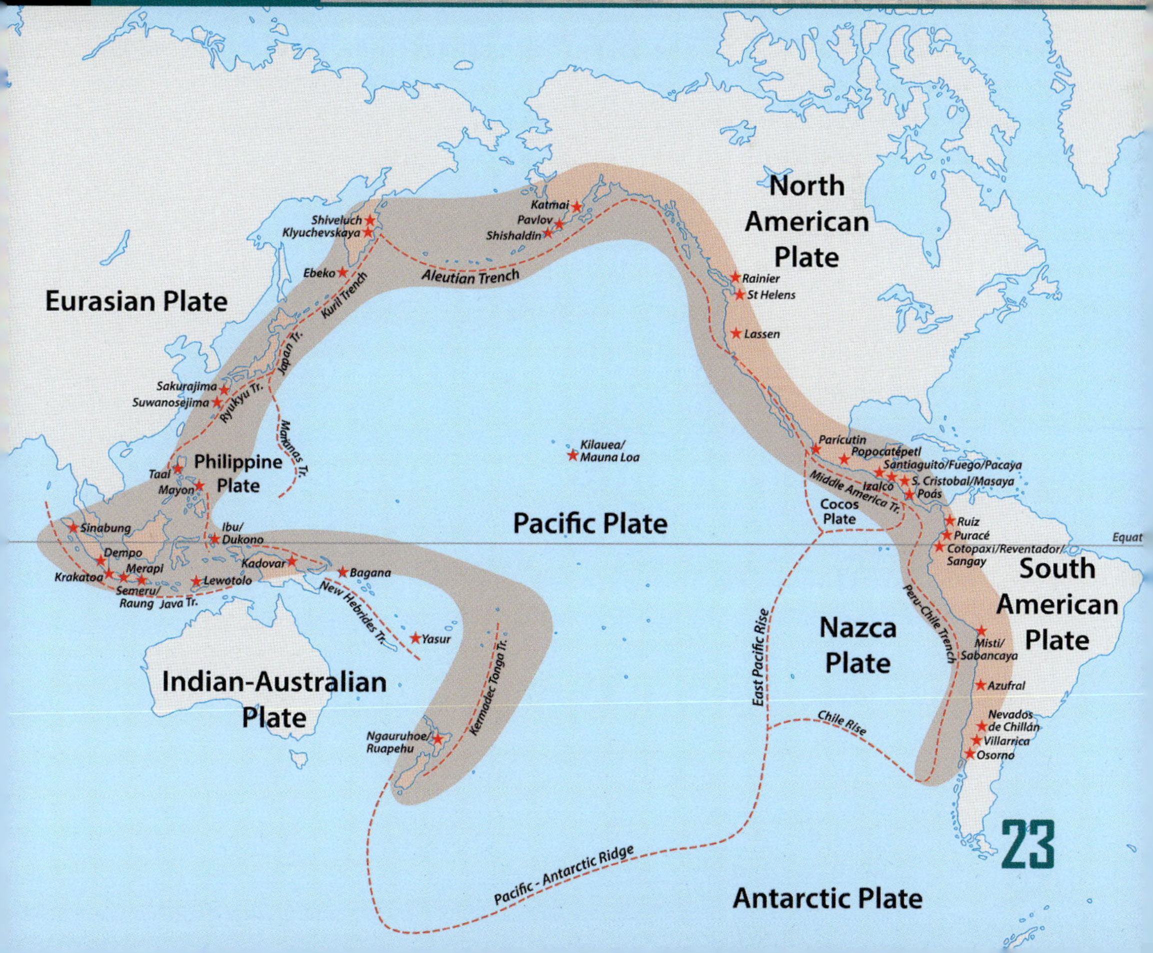

APPLIED GEOLOGY

Applied geology is using the study of geology to help people. For example, some geologists find useful elements and minerals inside Earth. They know where they're most likely to form because of their knowledge of Earth's processes. Many important tools, such as computers, wouldn't work without certain elements and minerals.

MAKE THE GRADE

A computer contains more than 30 elements and minerals, including gold, silica, nickel, aluminum, zinc, and iron. Geologists help companies and governments find this matter.

Applied geology also helps with construction, or building. Geologists can figure out if ground is stable, or firm, to support a heavy structure. They can also figure out if nearby mountains and hills are more likely to have landslides that could put people and buildings in danger.

MAKE THE GRADE

A landslide is a sudden movement of rocks and dirt down a hill or mountain. Rainfall, earthquakes, and other events can cause them.

GEOLOGY FOR THE FUTURE

Geology helps us understand our world's past as well as its present. And knowing about these can help us prepare for the future. Geology is a science that anyone can practice. Evidence of Earth's amazing history is all around us!

MAKE THE GRADE

Some geologists are working with other scientists to understand climate change. They've discovered today's **global warming** is happening much faster than any warming period of the past.

COMMON TOOLS OF GEOLOGISTS

rock hammer
a hammer used for breaking rocks

hand lens
a small, curved piece of glass used to make something look bigger

notebook
a place to record facts about findings and places

pencil magnet
a small magnet that helps reveal what a mineral is made of

compass
a tool that shows direction

camera
a recording tool used to explore areas or objects

sample bags
a way to collect bits of matter to bring back to a lab for study

GLOSSARY

climate: The average weather conditions of a place over a period of time.

continent: One of Earth's seven great landmasses.

cycle: A sequence of events that repeats.

earthquake: A shaking of the ground caused by the movement of Earth's crust.

element: Matter, such as gold, that is pure and has no other type of matter in it.

erosion: The act of wearing away by water, wind, or ice.

global warming: An increase in average global temperature.

layer: One thickness of something lying over or under another.

mineral: Matter in the ground that forms rocks.

pressure: A force that pushes on something else.

process: A series of steps or actions to complete something.

reptile: An animal covered with scales or plates that breathes air, has a backbone, and lays eggs.

weathering: Changing in appearance because of weather such as rain, cold, or wind.

FOR MORE INFORMATION

Books

Dodd, Emily. *Geology: Science Is All Around You!* London, England: Kingfisher, 2022.

Olson, Elsie. *Geology Lab: Explore Earth with Art and Activities.* Minneapolis, MN: Abdo Publishing, 2024.

Website

How to Look at a Rock Like a Geologist
www.thoughtco.com/how-to-look-at-a-rock-1441184
Find out how to read the clues in rocks yourself.

Publisher's note to educators and parents: Our editors have carefully reviewed these websites to ensure that they are suitable for students. Many websites change frequently, however, and we cannot guarantee that a site's future contents will continue to meet our high standards of quality and educational value. Be advised that students should be closely supervised whenever they access the internet.

INDEX